WE IMAGINE

Wade Hobbs

Copyright, 2014.

Wade Hobbs

ISBN-13: 978-1500512187

ISBN-10: 1500512184

To the People at NASA Who Build
Paper Airplanes

The Pleiades star cluster. Image courtesy NASA
and Antonio Fernandez-Sanchez.

Pleiades

I look to the east. The Pleiades appears almost like a comet but she is quite different. A group of stars, the Pleiades shines brilliantly in the Northern Hemisphere in December. And we see different things according to the instruments we use, if we use any at all. The Pleiades is a hazy V-shape formation, high in the sky.

I saw the Pleiades, and she appeared in the heavens with splendor. I could see her. She offered an escape, and she offered help. Greek in name only, she had a different fate, viewed from America... Her sisters loved her and they shared her fate...

For many, she offers hope. She tells of a different way of thinking. She beckons into a different tomorrow, and people stand on earth... and they wonder when.

Readers can start at the library. I hope they will not stop there in their search for the Pleiades. On a clear night, away from the city, it is easy enough to look up and see her. I hope readers find her.

I see her now, the Pleiades. It is seven o'clock, and she is high above, but I cannot follow her. It is too cold...

George Berkeley

I don't understand why people skip George Berkeley. People should read him. He's worth it. He begins simply.

He must surely be either very weak, or very little acquainted with the sciences, who shall reject a truth that is capable of demonstration, for no other reason but because it is newly known, and contrary to the prejudices of mankind. Thus much I thought fit to premise, in order to prevent, if possible, the hasty censures of a sort of men who are too apt to condemn an opinion before they rightly comprehend it. (A Treatise Concerning the Principles of Human Knowledge, George Berkeley)

He's often remembered as an empiricist. Berkeley did a lot for astronomy and thinking generally. People don't appreciate that. But he wrote many things. I wouldn't classify him as strictly a physicist. He wrote,

The only thing whose existence we deny IS

THAT WHICH PHILOSOPHERS CALL MATTER... (A Treatise Concerning the Principles of Human Knowledge, by George Berkeley)

I would hardly call that "empirical." It's more mind-over-matter.

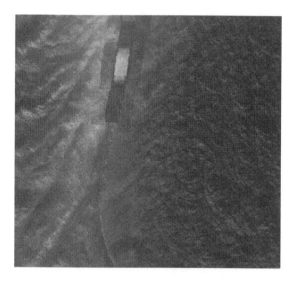

Infrared images of Enceladus' fissures.
Image courtesy of NASA and Cassini Team.

Newton was the ultimate empiricist. A ball drops. It hits the ground every time. That's empiricism.

I love astronomy because it brings freedom. We look. We listen.

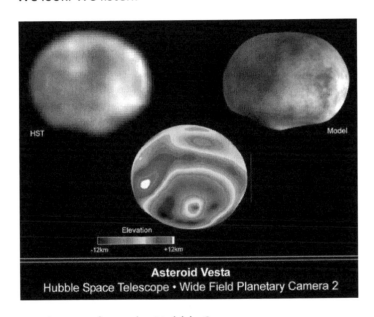

Asteroid Vesta
Hubble Space Telescope • Wide Field Planetary Camera 2

Two images from the Hubble Space Telescope. The first is a photo (top left). The second is a topographical photo created with radar (bottom). The model is computer generated. Image courtesy NASA Hubble Space Telescope Team.

We design a lens to find something, and we often find it. Readers can imagine! I think of infrared telescopes. We understand heat. So we build instruments that perpetuate our understanding of the world. Infrared telescopes show heat.

We don't build instruments to observe phenomenon we don't understand. I believe Berkeley's thinking applies particularly to astronomy and microbiology. In both, we design instruments to detect things we first create in our minds.

In astronomy, a wonderful world often emerges. We create an instrument. We hope to find something. Often, we don't. But sometimes we do.

Astronomy frees the mind.

Writing after Newton, Berkeley looked. He wondered. On the east coast of the United States, we still wonder. Just as Berkeley stood on America's shores some three hundred years ago, we stand and look into the heavens. And what did Berkeley write?

...all the unthinking objects of the mind agree in that they are entirely passive, and their existence consists only in being perceived... (A Treatise Concerning the Principles of Human Knowledge, George Berkeley)

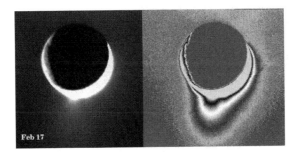

Above: Two images of Saturn's moon

Enceladus. On right is an infrared photo. It shows the icy plumes from the moon's South Pole. Image courtesy NASA and Cassini Team.

Most of us don't scrutinize things. It may seem trivial but it's significant. Berkeley debated Newton's theory of gravity. His critique paved the way for Albert Einstein's theories. We now know that Einstein's theory of relativity better explains Mercury's orbit, among other things.

ATTRACTION SIGNIFIES THE EFFECT, NOT THE MANNER OR CAUSE.--The great mechanical principle now in vogue is attraction. That a stone falls to the earth, or the sea swells towards the moon, may to some appear sufficiently explained thereby. But how are we enlightened by being told this is done by attraction? Is it that that word signifies the manner of the tendency, and that it is by the mutual drawing of bodies instead of their being impelled or protruded towards each other? But, nothing is determined of the manner or action, and it may as truly (for

aught we know) be termed "impulse," or "protrusion," as "attraction." Again, the parts of steel we see cohere firmly together, and this also is accounted for by attraction; but, in this as in the other instances, I do not perceive that anything is signified besides the effect itself; for as to the manner of the action whereby it is produced, or the cause which produces it, these are not so much as aimed at. (A Treatise Concerning the Principles of Human Knowledge, Berkeley)

Here, Berkeley criticizes Newton's theory of gravity. But even today, Newton's theory is useful. It explains a ball dropping to the ground. Rocket engineers still use it for moon trips and satellite launches.

But Newton's theory did not explain Mercury's orbit. Slight inconsistencies plagued the theory. Einstein's general relativity theory finally explained the orbit of Mercury. So Berkeley laid the groundwork for later advances in science.

A wise man once said, "We see what we want to see, and we hear what we want to hear." This diagram shows the Wilkinson Probe that mapped the background radiation from the Big Bang. The Wilkinson Team created the Wilkinson Microwave Anisotropy Map. Image courtesy NASA and the Wilkinson Team.

Berkeley continues:

...by a diligent observation of the phenomena within our view, we may discover the general laws of nature, and from them deduce the other phenomena; I

16

do not say demonstrate, for all deductions of that kind depend on a supposition that ...nature always operates uniformly, and in a constant observance of those rules we take for principles: which we cannot evidently know. (A Treatise Concerning the Principles of Human Knowledge, George Berkeley)

Mariner 5. Our probes are designed to find things we first understand in the mind. Image courtesy NASA and Mariner Team.

So he recognizes scientific method. But he questions its universality. Often we find situations that science doesn't explain. Physicists, for example, cannot explain what happens inside a black hole.

Berkeley continues:

110. The best key for the aforesaid analogy or natural Science will be easily acknowledged to be a certain celebrated Treatise of Mechanics. In the entrance of which justly admired treatise, Time, Space, and Motion are distinguished into absolute and relative, true and apparent, mathematical and vulgar; which distinction, as it is at large explained by the author, does suppose these quantities to have an existence without the mind; and that they are ordinarily conceived with relation to sensible things, to which nevertheless in their own nature they bear no relation at all.

111. As for Time, as it is there taken in an absolute or abstracted sense, for the duration or perseverance of the existence of things, I have nothing more to add concerning it after what has been already

said on that subject... For the rest, this celebrated author holds there is an absolute Space, which, being unperceivable to sense, remains in itself similar and immovable; and relative space to be the measure thereof, which, being movable and defined by its situation in respect of sensible bodies, is vulgarly taken for immovable space. Place he defines to be that part of space which is occupied by any body; and according as the space is absolute or relative so also is the place. Absolute Motion is said to be the translation of a body from absolute place to absolute place, as relative motion is from one relative place to another. And, because the parts of absolute space do not fall under our senses, instead of them we are obliged to use their sensible measures, and so define both place and motion with respect to bodies which we regard as immovable. But, it is said in philosophical matters we must abstract from our senses, since it may be that none of those bodies which seem to be quiescent are truly so, and the same thing which is moved relatively may be really at rest; as likewise one and the same body may be in relative rest and motion, or even moved with contrary relative motions at the same time, according as its place is variously defined. All which ambiguity is to be found in the apparent motions, but not at all in the true or absolute, which should therefore be alone regarded in philosophy. And the true as we are told are distinguished from apparent or relative motions by the following properties.--First, in true or absolute motion all parts which preserve the same position with respect of the whole, partake of the motions of the whole. Secondly, the place being moved, that which is placed therein is also moved; so that a body moving in a place which is in motion doth participate the motion of its place. Thirdly, true motion is never generated or changed otherwise than by force impressed on the body itself. Fourthly, true motion is always changed by force impressed on the body moved. Fifthly, in circular motion barely relative there is no centrifugal force, which, nevertheless, in that which is true or absolute, is proportional to the quantity of motion.

112. MOTION, WHETHER REAL OR APPARENT, RELATIVE.--But, notwithstanding what has been said, I must confess it does not appear to me that there can be any motion other than relative; so that to conceive motion there must be at least conceived two bodies, whereof the distance or position in regard to each other is varied. Hence, if there was one only body in being it could not possibly be moved. This seems evident, in that the idea I have of motion doth necessarily include relation. (A Treatise Concerning the Principles of Human Knowledge, George Berkeley)

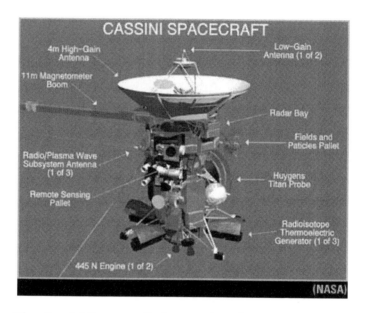

The Cassini Spacecraft that explored Saturn and Titan. Image courtesy NASA and The Cassini Team.

Today, physicists commonly write that the idea of absolute time and space has been rejected. Einstein disproved absolute time and space. In fact, the seeds of relative time and space can be traced back to Berkeley, if not to the Ancient Greeks. Einstein's Theory of Relativity replaces absolute time and space with space-time.

According to Einstein, time and space are relative.

Einstein

Everyone should appreciate Einstein. He was tremendously skeptical, and we would all do well to approach every theory with his skepticism. His analysis of Euclid's geometry is worth every precious word. And beyond science, he was a leading pacifist.

Many revere Einstein as the smartest man in history. I won't argue. Readers should not bother coordinating moving clocks.

In Einstein, readers can find hope and skepticism. That's the way I see it. Gravity waves are part of the picture.

Gravity waves imply big trebles in space-time. We have confirmed gravity waves at L.S.U., at L.I.G.O. That's Louisiana State University, and Laser Interferometer

Gravitational-wave Observatory.

In my opinion, gravity waves may explain many of astronomy's inconsistencies. An American observatory records one thing in 2014. Astronomers check Chinese and Arab observations from centuries ago, and find differences in all the observations. It may be that a giant gravity wave moved a stellar phenomenon.

We know that gravity waves exist. Astrophysicists observed them. But to the best of my knowledge, we haven't seen a major gravity wave moving a stellar phenomenon. That's just one explanation.

But Einstein's thinking implies that a huge gravity wave might move a stellar phenomenon.

Minkowski

17. MINKOWSKI'S FOUR-DIMENSIONAL SPACE

The non-mathematician is seized by a mysterious shuddering when he hears of "four-dimensional" things, by a feeling not unlike that awakened by thoughts of the occult. And yet there is no more common-place statement than that the world in which we live is a four-dimensional space-time continuum. (Einstein, Theory of Relativity)

With all due respect to The Great Thinker Albert Einstein, I can think of a more common-place statement than "the world in which we live is a four-dimensional space-time continuum." Let me think of an example. "The apple falls to earth." Sometimes Einstein is so counterintuitive I simply break for coffee.

Readers who find Einstein intuitive can work on satellite communications. Relativity applies.

When Einstein writes of "the occult," he refers to cutting-edge physics. That's why I love to read him. New ideas are always interesting. We are still learning the applications of Einstein's thought.

The Magellan. Image courtesy NASA and the Magellan Team.

A Bit Ironic

You may find it a bit ironic. You may find it kind of weird. But a place called Berkeley leads astronomy.

The only thing whose existence we deny IS THAT WHICH PHILOSOPHERS CALL MATTER... (A Treatise Concerning the Principles of Human Knowledge, by George Berkeley)

Here is Berkeley the philosopher writing. Isn't it weird that a science like astronomy is lead by a man who denied matter?

And if you don't believe Berkeley leads, consider the efforts of the University of Berkeley Team, lead by Professor Geoffrey Marcey. Professor Marcey and his team look through the skies for a new earth. They're making progress.

Skepticism

Berkeley himself was skeptical of Newton, so it's odd that he criticized skepticism. I suppose Berkeley was skeptical of skepticism. He wrote:

Philosophy being nothing else but THE STUDY OF WISDOM AND TRUTH, it may with reason be expected that those who have spent most time and pains in it should enjoy a greater calm and serenity of mind, a greater clearness and evidence of knowledge, and be less disturbed with doubts and difficulties than other men. Yet so it is, we see the illiterate bulk of mankind that walk the high-road of plain common sense, and are governed by the dictates of nature, for the most part easy and undisturbed. To them nothing THAT IS FAMILIAR appears unaccountable or difficult to comprehend. They complain not of any want of evidence in their senses, and are out of all danger of becoming SCEPTICS. But no sooner do we depart from sense and instinct to follow

the light of a superior principle, to reason, meditate, and reflect on the nature of things, but a thousand scruples spring up in our minds concerning those things which before we seemed fully to comprehend. Prejudices and errors of sense do from all parts discover themselves to our view; and, endeavoring to correct these by reason, we are insensibly drawn into uncouth paradoxes, difficulties, and inconsistencies, which multiply and grow upon us as we advance in speculation, till at length, having wandered through many intricate mazes, we find ourselves just where we were, or, which is worse, sit down in a forlorn Scepticism. (Berkeley, Treatise Concerning the Principles of Human Knowledge)

Einstein seems naturally skeptical. As history's leading physicist, he led in the search for uniform physical laws. He wrote:

...the most careful observations have never revealed... anisotropic properties in terrestrial physical space... (Albert

Einstein, Theory of Relativity)

The Galileo Mission explored Jupiter and its moons. The 6502 chip was used in Apple and Commodore 64-bit computers. It was also used to control this sophisticated probe! Electromagnetic observations confirmed Europa's ocean. Image courtesy of NASA, the Jet Propulsion Lab, and The Galileo Team.

Here he clearly articulates his philosophy of science:

From a systematic theoretical point of view, we may imagine the process of evolution of an empirical science to be a

31

continuous process of induction. Theories are evolved and are expressed in short compass as statements of a large number of individual observations in the form of empirical laws, from which the general laws can be ascertained by comparison. Regarded in this way, the development of a science bears some resemblance to the compilation of a classified catalogue. It is, as it were, a purely empirical enterprise.

But this point of view by no means embraces the whole of the actual process; for it slurs over the important part played by intuition and deductive thought in the development of an exact science. As soon as a science has emerged from its initial stages, theoretical advances are no longer achieved merely by a process of arrangement. Guided by empirical data, the investigator rather develops a system of thought which, in general, is built up logically from a small number of fundamental assumptions, the so-called axioms. We call such a system of thought a theory. The theory finds the justification for its existence in the fact that it correlates a large number of single observations, and it is just here that the "truth" of the theory lies. (Einstein, Theory of Relativity)

In another passage, Einstein adds:

No fairer destiny could be allotted to any physical theory, than that it should of itself point out the way to the introduction of a more comprehensive theory, in which it lives on as a limiting case. (Einstein, Theory of Relativity)

In my view, we formulate physics theories, and in certain contexts they work. In others, they don't. We have mechanics. We have electromagnetism. We have relativity.

I agree with Einstein that physics is fun. Obviously, we should continue to seek laws that explain the universe. But I believe we formulate templates that work in some places only to give way to other theories in different contexts.

Those who followed Newton in his day were so convinced that his theory was a tremendous work. It lasted less than three hundred years. Einstein's General Relativity is now accepted. I see continual supplanting of ideas and theories. Rather than "laws," physics is more a set of templates.

They work for a while, and then they are supplanted. As I see it, the universe is anisotropic.

I hope physicists don't accept that. We're having fun searching for nature's laws.

Newton's Response

Were Newton here he might respond to Berkeley. He might say, "They still use my math when they launch something into space. It's still a good estimate. And as Einstein recognized, my theories still explain the orbits of most of the planets."

Berkeley would be loathe to argue.

Newton might continue. "Drop a ball. It falls to the ground. That's what I predict. Figure that out."

Jupiter. Image courtesy NASA.

Again, Berkeley would not argue.

Newton might point to the effectiveness of calculus, which he invented. It's used every day.

Berkeley was quite specific on this point:

130. Of late the speculations about Infinities have run so high, and grown to such strange notions, as have occasioned no small scruples and disputes among the geometers of the present age. Some there are of great note who, not content with holding that finite lines may be divided into an infinite number of parts, do yet farther maintain that each of those infinitesimals is itself subdivisible into an infinity of other parts or infinitesimals of a second order, and so on ad infinitum. These, I say, assert there are infinitesimals of infinitesimals of infinitesimals, &c., without ever coming to an end; so that according to them an inch does not barely contain an infinite number of parts, but an infinity of an infinity of an infinity ad infinitum of parts. Others there be who hold all orders of infinitesimals below the first to be nothing at all; thinking it with good reason absurd to imagine there is any positive quantity or part of extension which, though multiplied infinitely, can never equal the smallest given extension. And yet on the other hand it seems no less

absurd to think the square, cube or other power of a positive real root, should itself be nothing at all; which they who hold infinitesimals of the first order, denying all of the subsequent orders, are obliged to maintain.

131. OBJECTION OF MATHEMATICIANS.-- ANSWER.--Have we not therefore reason to conclude they are both in the wrong, and that there is in effect no such thing as parts infinitely small, or an infinite number of parts contained in any finite quantity? But you will say that if this doctrine obtains it will follow the very foundations of Geometry are destroyed, and those great men who have raised that science to so astonishing a height, have been all the while building a castle in the air. To this it may be replied that whatever is useful in geometry, and promotes the benefit of human life, does still remain firm and unshaken on our principles; that science considered as practical will rather receive advantage than any prejudice from what has been said. But to set this in a due light may be the proper business of another place. For the rest, though it should follow

that some of the more intricate and subtle parts of Speculative Mathematics may be pared off without any prejudice to truth, yet I do not see what damage will be thence derived to mankind. On the contrary, I think it were highly to be wished that men of great abilities and obstinate application would draw off their thoughts from those amusements, and employ them in the study of such things as lie nearer the concerns of life, or have a more direct influence on the manners.

132. SECOND OBJECTION OF MATHEMATICIANS.--ANSWER.--If it be said that several theorems undoubtedly true are discovered by methods in which infinitesimals are made use of, which could never have been if their existence included a contradiction in it; I answer that upon a thorough examination it will not be found that in any instance it is necessary to make use of or conceive infinitesimal parts of finite lines, or even quantities less than the minimum sensible; nay, it will be evident this is never done, it being impossible. (Berkeley, A Treatise Concerning the Principles of Human Knowledge)

I suppose that, were Newton here, he'd say, "Dude. You paineth me greatly. People still use calculus."

Previous page: Different views of Jupiter's moon Europa. Image courtesy of NASA, the Jet Propulsion Lab, and the Galileo Team.

Avoiding War

The United States, Britain, and Canada are making a big mistake if they fail to explore Saturn and Jupiter. In this world there are nuts who constantly want war. Why they want war I'll never understand. As Roosevelt said, "War is hell."

Saturn's moon Titan holds vast amounts of methane. That can be used to propel rockets or heat dwellings. Another of Saturn's moons, Enceladus, holds a vast body of saltwater under an icy surface. My research proves that microbial life exists in that water. Larger life forms may live beneath Enceladus' surface.

Jupiter's moons hold the promise of larger life forms, too. Europa, Ganymede, and Callisto all hold subsurface saltwater. As I prove in my books, "OTHER LIFE EXISTS," "QUESTION ANSWERED," and "KNOWN," those moons must hold microscopic life. By my estimates, a good chance exists that larger life forms may live in the waters of at least one of those moons. We should search for life in all of them. Aquatic life there represents a food source for later generations.

As Carl Sagan wrote, nuclear war on this planet could be a disaster. World War II was bad enough. War came again on September 11, 2001. I hate it. In my opinion, the people who flee war and peacefully explore other planets are perfect.

We are always justified in saving lives.

The Brits are among the most capable people in the world. I show favoritism. I know that if they decide to drop broccoli from the sky, they can. And it's much easier to grow broccoli above earth and drop it than it is to go back to Babylon for oil.

America, Britain and Canada don't need Middle Eastern oil. How many times have we raced Germany, Russia, China, and every other country on this planet back to the Middle East for oil? Countless times!

Perhaps it's a bit optimistic for the year 2014. But I believe we can grow broccoli in greenhouses above earth and drop it. It is critical in my view to reward innovation. Using gravity to drop food from 300 miles above earth will be easier than moving it across the United States or Canada. As oil prices rise, greenhouses above earth become a more practical alternative.

In my opinion, greenhouses are more consistent with the Apollo philosophy, "We came in peace for all mankind."

America's Aldrin on the moon. Image courtesy of NASA.

Einstein's Skepticism

In your schooldays most of you who read this book made acquaintance with the noble building of Euclid's geometry, and you remember--perhaps with more respect than love--the magnificent structure, on the lofty staircase of which you were chased about for uncounted hours by conscientious teachers. By reason of our past experience, you would certainly regard everyone with disdain who should pronounce even the most out-of-the-way proposition of this science to be untrue. But perhaps this feeling of proud certainty would leave you immediately if some one were to ask you: "What, then, do you mean by the assertion that these propositions are true?" Let us proceed to give this question a little consideration. (Einstein, Theory of Relativity, 1916)

Beyond the substance of Einstein's work, I note that he brings an unusual skepticism to his subject. He approaches a standard school of thought, Euclid's geometry, and analyzes it with careful detail. He invites us to the road to broader truths when he writes, **"What, then, do you mean by the assertion that these propositions are true?"** He accepts nothing at face value.

Geometry sets out from certain conceptions such as "plane," "point," and "straight line," with which we are able to associate more or less definite ideas, and from certain simple propositions (axioms) which, in virtue of these ideas, we are inclined to accept as "true." Then, on the basis of a logical process, the justification of which we feel ourselves compelled to admit, all remaining propositions are shown to follow from those axioms, i.e. they are proven. A proposition is then correct ("true") when it has been derived in the recognised manner from the axioms. The question of "truth" of the individual geometrical propositions is thus reduced to one of the "truth" of the axioms. Now it has long been known that the last question is not only unanswerable by the methods of geometry, but that it is in itself entirely without meaning. We cannot ask whether it is true that only one straight line goes through two points. We can only say that Euclidean geometry deals with things called "straight lines," to each of which is ascribed the property of being uniquely determined by two points situated on it. The concept "true" does not tally with the assertions of pure geometry, because by the word "true" we are eventually in the habit of designating always the correspondence with a "real" object; geometry, however, is not concerned with the relation of the ideas involved in it to objects of experience, but only with the logical connection of these ideas among themselves. (Einstein, Theory of Relativity, 1916)

Einstein continues his deft analysis. My point is that Einstein brought untold insight to his subject. He questioned every aspect of Euclid's thought. His method is even more important than the substance of his theories.

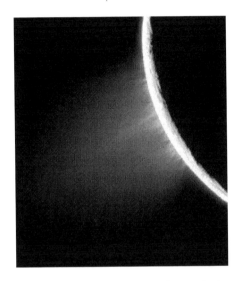

The promising plumes of Enceladus. Image courtesy NASA and Cassini Team.

Einstein limits the "truth" of the axioms of Euclid's geometry to a truth within that intellectual framework. He doesn't seek to "disprove" Euclid. Rather, he simple qualifies him.

It is not difficult to understand why, in spite of this, we feel constrained to call the propositions of geometry "true."

Geometrical ideas correspond to more or less exact objects in nature, and these last are undoubtedly the exclusive cause of the genesis of those ideas. Geometry ought to refrain from such a course, in order to give to its structure the largest possible logical unity. The practice, for example, of seeing in a "distance" two marked positions on a practically rigid body is something which is lodged deeply in our habit of thought. We are accustomed further to regard three points as being situated on a straight line, if their apparent positions can be made to coincide for observation with one eye, under suitable choice of our place of observation. (Einstein, Theory of Relativity, 1916)

On this intellectual journey with Einstein, we encounter only habits of thought. Euclid's geometry and Newton's theory of gravity are such habits of thought. We must review every idea, regardless of how widely accepted, and analyze its truth. The "laws" of physics and mathematics become mere habits of thought subject to scrutiny.

If, in pursuance of our habit of thought, we now supplement the propositions of Euclidean geometry by the single proposition that two points on a practically rigid body always correspond to the same distance (line-interval), independently of any changes in position to which we may subject the body, the propositions of Euclidean geometry then resolve themselves into propositions on the possible relative position of practically rigid bodies. Geometry which has been supplemented in this way is then to be treated as a branch of physics. We can now legitimately ask as to the "truth" of geometrical propositions interpreted in this way, since we are justified in asking whether these propositions are satisfied for those real things we have associated with the geometrical ideas. In less exact terms we can express this by saying that by the "truth" of a geometrical proposition in this sense we understand its validity for a construction with rule and compasses.

Of course the conviction of the "truth" of

geometrical propositions in this sense is founded exclusively on rather incomplete experience. For the present we shall assume the "truth" of the geometrical propositions, then at a later stage (in the general theory of relativity) we shall see that this "truth" is limited, and we shall consider the extent of its limitation. (Einstein, Theory of Relativity, 1916)

I'll continue with more Einstein because his method is priceless. When we approach astrobiology with his skepticism, we can answer a fundamental question. But before we approach the substance of astrobiology, we should learn skepticism from a master.

THE SYSTEM OF CO-ORDINATES

On the basis of the physical interpretation of distance which has been indicated, we are also in a position to establish the distance between two points on a rigid body by means of measurements. For this purpose we require a "distance" (rod S) which is to be used once and for all, and which we employ as a standard measure. If, now, A and B are two points on a rigid body, we can construct the line joining them according to the rules of geometry ; then, starting from A, we can mark off the distance S time after time until we reach B. The number of these operations required is the numerical measure of the distance AB. This is the basis of all measurement of length.

Every description of the scene of an event or of the position of an object in space is based on the specification of the point on a rigid body (body of reference) with which that event or object coincides. This applies not only to scientific description, but also to everyday life. If I analyse the place specification "Trafalgar Square, London," I arrive at the following result. The earth is the rigid body to which the specification of place refers; "Trafalgar Square, London," is a well-defined point, to which a name has been assigned, and with which the event coincides in space. (Einstein, Theory of Relativity, 1916)

Americans partying on moon. Image
courtesy of NASA.

Here Einstein reviews basic geometry. He
continues.

**This primitive method of place
specification deals only with places on the
surface of rigid bodies, and is dependent
on the existence of points on this surface
which are distinguishable from each other.
But we can free ourselves from both of**

these limitations without altering the nature of our specification of position. If, for instance, a cloud is hovering over Trafalgar Square, then we can determine its position relative to the surface of the earth by erecting a pole perpendicularly on the Square, so that it reaches the cloud. The length of the pole measured with the standard measuring-rod, combined with the specification of the position of the foot of the pole, supplies us with a complete place specification. On the basis of this illustration, we are able to see the manner in which a refinement of the conception of position has been developed.

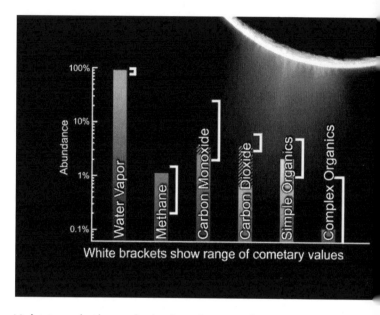

White brackets show range of cometary values

Light travels through the icy plumes of Enceladus. According to Einstein, light travels at a constant speed **c** through the vacuum. The light slows as it passes through the plume material. We use spectrography to analyze the chemistry of comets and Enceladus' plumes. Image courtesy of NASA.

(a) We imagine the rigid body, to which the place specification is referred, supplemented in such a manner that the object whose position we require is reached by the completed rigid body.

(b) In locating the position of the object, we make use of a number (here the length of the pole measured with the measuring-rod) instead of designated points of reference.

(c) We speak of the height of the cloud even when the pole which reaches the cloud has not been erected. By means of optical observations of the cloud from different positions on the ground, and taking into account the properties of the propagation of light, we determine the length of the pole we should have required in order to reach the cloud. (Einstein, Theory of Relativity, 1916)

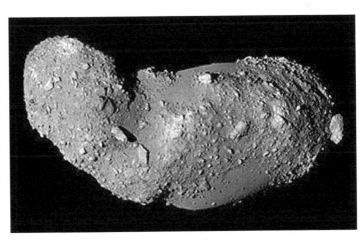

This asteroid could be either the stationary body or the moving body in Einstein's theory. Image courtesy NASA.

Einstein proceeds with caution, not violating Euclidean geometry, but not becoming paralyzed by it. He explains the connection with astronomy.

From this consideration we see that it will be advantageous if, in the description of position, it should be possible by means of numerical measures to make ourselves independent of the existence of marked positions (possessing names) on the rigid body of reference. In the physics of measurement this is attained by the application of the Cartesian system of co-ordinates.

This consists of three plane surfaces perpendicular to each other and rigidly attached to a rigid body. Referred to a system of co-ordinates, the scene of any event will be determined (for the main part) by the specification of the lengths of

the three perpendiculars or co-ordinates (x, y, z) which can be dropped from the scene of the event to those three plane surfaces. The lengths of these three perpendiculars can be determined by a series of manipulations with rigid measuring-rods performed according to the rules and methods laid down by Euclidean geometry. (Einstein, Theory of Relativity, 1916)

Measuring-rods play a critical role in Einstein's theory. Later, we visit the concept of "length contraction."

In practice, the rigid surfaces which constitute the system of co-ordinates are generally not available; furthermore, the magnitudes of the co-ordinates are not actually determined by constructions with rigid rods, but by indirect means. If the results of physics and astronomy are to maintain their clearness, the physical meaning of specifications of position must always be sought in accordance with the above considerations.

We thus obtain the following result: Every description of events in space involves the use of a rigid body to which such events have to be referred. The resulting relationship takes for granted that the laws of Euclidean geometry hold for "distances;" the "distance" being represented physically by means of the convention of two marks on a rigid body. (Einstein, Theory of Relativity, 1916)

Next, Einstein presents a standard example. In a moving railway carriage, an observer in the train sees things differently from a stationary observer on the ground. Einstein unravels the classic notions of absolute space and time.

3. SPACE AND TIME IN CLASSICAL MECHANICS

The purpose of mechanics is to describe

how bodies change their position in space with "time." I should load my conscience with grave sins against the sacred spirit of lucidity were I to formulate the aims of mechanics in this way, without serious reflection and detailed explanations. Let us proceed to disclose these sins.

It is not clear what is to be understood here by "position" and "space." I stand at the window of a railway carriage which is travelling uniformly, and drop a stone on the embankment, without throwing it. Then, disregarding the influence of the air resistance, I see the stone descend in a straight line. A pedestrian who observes the misdeed from the footpath notices that the stone falls to earth in a parabolic curve. I now ask: Do the "positions" traversed by the stone lie "in reality" on a straight line or on a parabola? Moreover, what is meant here by motion "in space"? From the considerations of the previous section the answer is self-evident. In the first place we entirely shun the vague word "space," of which, we must honestly acknowledge, we cannot form the slightest conception, and we replace it by "motion relative to a practically rigid body of reference." The positions relative to the body of reference (railway carriage or embankment) have already been defined in detail in the preceding section. If instead of "body of reference" we insert "system of co-ordinates," which is a useful idea for mathematical description, we are in a position to say : The stone traverses a straight line relative to a system of co-ordinates rigidly attached to the carriage, but relative to a system of co-ordinates rigidly attached to the ground (embankment) it describes a parabola. With the aid of this example it is clearly seen that there is no such thing as an independently existing trajectory (lit. "path-curve"), but only a trajectory relative to a particular body of reference. (Einstein, Theory of Relativity)

Enceladus and two other moons of Saturn.
Image courtesy of NASA.

As a patent examiner, Einstein was familiar
with a technical problem: Efforts to
coordinate clocks between moving trains
and their stations. Einstein next introduces
one of two basic rules of relativity. Light
travels at a constant velocity c through a
vacuum.

**In order to have a complete description of
the motion, we must specify how the body
alters its position with time ; i.e. for every
point on the trajectory it must be stated at**

what time the body is situated there. These data must be supplemented by such a definition of time that, in virtue of this definition, these time-values can be regarded essentially as magnitudes (results of measurements) capable of observation. If we take our stand on the ground of classical mechanics, we can satisfy this requirement for our illustration in the following manner. We imagine two clocks of identical construction; the man at the railway-carriage window is holding one of them, and the man on the footpath the other. Each of the observers determines the position on his own reference-body occupied by the stone at each tick of the clock he is holding in his hand. In this connection we have not taken account of the inaccuracy involved by the finiteness of the velocity of propagation of light. With this and with a second difficulty prevailing here we shall have to deal in detail later. (Einstein, Theory of Relativity, 1916)

After reviewing the Galileian system of coordinates, Einstein presents the theory of special relativity. He would work another decade refining the theory into general relativity.

5. THE PRINCIPLE OF RELATIVITY (IN THE RESTRICTED SENSE)

In order to attain the greatest possible clearness, let us return to our example of the railway carriage supposed to be travelling uniformly. We call its motion a uniform translation ("uniform" because it is of constant velocity and direction, "translation" because although the carriage changes its position relative to the embankment yet it does not rotate in so doing). Let us imagine a raven flying through the air in such a manner that its motion, as observed from the embankment, is uniform and in a straight line. If we were to observe the flying raven from the moving railway carriage. we should find that the motion of the raven would be one of different velocity and direction, but that it would still be uniform and in a straight line. Expressed in an abstract manner we may say : If a mass m is moving uniformly in a straight line with respect to a co-ordinate system K, then it will also be moving uniformly and in a straight line relative to a second co-ordinate system K' provided that the latter is executing a uniform translatory motion with respect to K. In accordance with the discussion contained in the preceding section, it follows that:

If K is a Galileian co-ordinate system. then every other co-ordinate system K' is a Galileian one, when, in relation to K, it is in a condition of uniform motion of translation. Relative to K' the mechanical laws of Galilei-Newton hold good exactly as they do with respect to K. (Einstein, Theory of Relativity, 1916)

Here, Einstein discloses the second principle of relativity: The mechanical laws hold true in a moving frame of reference just as they do in a stationary reference frame. More generally, the laws of physics hold true in moving frames of reference. A frame of reference is simply a three dimensional co-ordinate system Einstein describes.

Again, my object is not the substance of the theory of relativity, though the theory is quite interesting in itself. Rather, I'm concerned with Einstein's meticulous reasoning. We will later employ his careful reasoning style in proving a point about astrobiology.

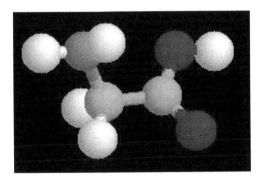

Einstein's meticulous reasoning technique is useful in astrobiology. Above: Amino acid found by the Stardust Mission. Image courtesy NASA.

We advance a step farther in our generalisation when we express the tenet thus: If, relative to K, K' is a uniformly moving co-ordinate system devoid of rotation, then natural phenomena run their course with respect to K' according to exactly the same general laws as with respect to K. This statement is called the principle of relativity (in the restricted sense)...

... there are two general facts which at the outset speak very much in favour of the validity of the principle of relativity. Even though classical mechanics does not supply us with a sufficiently broad basis for the theoretical presentation of all physical phenomena, still we must grant it a considerable measure of "truth," since it supplies us with the actual motions of the heavenly bodies with a delicacy of detail little short of wonderful. The principle of relativity must therefore apply with great accuracy in the domain of mechanics. But that a principle of such broad generality should hold with such exactness in one domain of phenomena, and yet should be invalid for another, is a priori not very probable. (Einstein, Theory of Relativity, 1916)

We later find that classical mechanics does not explain the perihelion of Mercury. In fact, Einstein's general theory of relativity explains the perihelion of Mercury, and astronomical observations confirmed the theory.

We now proceed to the second argument, to which, moreover, we shall return later. If the principle of relativity (in the restricted sense) does not hold, then the Galileian co-ordinate systems K, K', K", etc., which are moving uniformly relative to each other, will not be equivalent for the description of natural phenomena. In this case we should be constrained to believe that natural laws are capable of being formulated in a particularly simple manner, and of course only on condition that, from amongst all possible Galileian co-ordinate systems, we should have chosen one (K0) of a particular state of motion as our body of reference. We should then be justified (because of its merits for the description of natural

phenomena) in calling this system "absolutely at rest," and all other Galileian systems K "in motion." If, for instance, our embankment were the system K0 then our railway carriage would be a system K, relative to which less simple laws would hold than with respect to K0. This diminished simplicity would be due to the fact that the carriage K would be in motion (i.e."really") with respect to K0. In the general laws of nature which have been formulated with reference to K, the magnitude and direction of the velocity of the carriage would necessarily play a part. We should expect, for instance, that the note emitted by an organpipe placed with its axis parallel to the direction of travel would be different from that emitted if the axis of the pipe were placed perpendicular to this direction.

Now in virtue of its motion in an orbit round the sun, our earth is comparable with a railway carriage travelling with a velocity of about 30 kilometers per second. If the principle of relativity were not valid we should therefore expect that the direction of motion of the earth at any moment would enter into the laws of nature, and also that physical systems in their behaviour would be dependent on the orientation in space with respect to the earth. For owing to the alteration in direction of the velocity of revolution of the earth in the course of a year, the earth cannot be at rest relative to the hypothetical system K0 throughout the whole year. However, the most careful observations have never revealed such anisotropic properties in terrestrial physical space, i.e. a physical non-equivalence of different directions. This is very powerful argument in favour of the principle of relativity. (Einstein, Theory of Relativity, 1916)

Sadly, Einstein passed away in 1955. Before he left, he defined Twentieth Century physics. He won the Nobel for the photoelectric effect. Of course, he postulated the theory of relativity. Schwarzschild used that theory to mathematically prove singularities, or black holes.

That's not all Einstein did. As biographer Walter Isaacson notes, Einstein co-authored a 1935 paper with Nathan Rosen and Boris Podosky entitled, "Can the Quantum-Mechanical Description of Physical Reality Be Regarded as Complete?"

A clock on earth counts time faster than a clock on a spaceship moving past her. Image courtesy NASA.

Among other things, Einstein solved a couple of problems. First, he knew of the problem of coordinating clocks between a train and its platform. Relativity proves that clocks in motion measure time at different rates than stationary clocks. Relativity proves time dilation and the twin paradox.

Secondly, astronomers debated the orbit of Mercury. Newton's theory of gravity failed to explain it. The theory of relativity explained it and astronomical observations confirmed the theory.

But Einstein did not consider all physical problems. In his day, scientists still debated the existence of black holes. To my knowledge, astronomers observed black holes only after Einstein's passing. They certainly never observed a black hole moving through space-time.

In fact, I know of no such observations even today. So that's one scenario Einstein didn't consider. I'll restate it for clarity. Einstein never considered the scenario of a black hole moving through space-time. In my opinion, this scenario presents a new challenge to physicists.

We know that relativity does not apply inside a black hole. Indeed, all the known laws of physics do not apply. That's how little we know about the inner workings of black holes. So this is a fruitful area for new physics.

Length Contraction and Time Dilation

Special relativity results in length contraction and time dilation. What does that mean?

First, let's look at length contraction. It's simple. A moving body holds a rod. The faster the body moves, the more the length of the rod contracts as measured by a stationary observer.

Let's think of ourselves measuring a ruler positioned inside a car. We stand outside the car. When the car is stationary, the ruler measures twelve inches. But when the car moves past at a high speed, our measurement is necessarily less than twelve inches. That's length contraction.

Time dilation is fabulous. Let's think of a clock inside a spaceship. We are stationary on earth. When the ship is stationary on earth, the ship's clock runs at the same pace as our ground clock. Then the ship launches.

When the ship passes earth at high speeds, time measurements on earth differ from the ship's clock. Time on the ship passes more slowly than time on earth. The ship's clock moves slower than the ground clock.

Americans partying on the moon. Image

courtesy of NASA.

Time dilation results in the twin paradox. A twin traveling at very high speeds on a spaceship ages slower than the twin who remains on earth. In practice, mass increases as velocity approaches the speed of light, so it's presently impractical to move a human at such speeds. But theoretically, the twin paradox exists. Time dilation implies it.

Chinese Takeout

Einstein's theory of relativity presents critical issues. Most importantly, if one is moving at near the speed of light, how long does it take to slow down for Chinese food?

Alternatively, presume one is traveling at relativistic speed, a velocity near the speed of light. Somebody on the spaceship orders pizza. Is delivery time calculated according to time on the ship, or according to the clock at the restaurant on earth?

It Wouldn't Surprise Me

It wouldn't surprise me if Asians are
currently flying little chip devices around
Saturn or Jupiter. Many things don't
surprise me.

It wouldn't surprise me if Europe has
reached Alpha Centauri by now. It wouldn't
surprise me if one of the Mediterranean
nations has explored new regions of space.
It wouldn't surprise me if Canada is past
Pluto. It wouldn't surprise me if Ireland is
doing something phenomenal in space.
Kennedy led the American space
movement.

South Africa controls gold. Who knows
what technology they've developed? It
wouldn't surprise me.

When the English reached the South Pacific,

they found plenty of surprises. They found new nations and islands. Among other things, they found Australia. Who knows what the Australians are doing? It wouldn't surprise me. Australia survived a bus-size piece of NASA junk that crashed in the outback.

We Imagine

We imagine a large portion of empty space, so far removed from stars and other appreciable masses, that we have before us approximately the conditions required by the fundamental law of Galilei. It is then possible to choose a Galileian reference-body for this part of space (world), relative to which points at rest remain at rest and points in motion continue permanently in uniform rectilinear motion. As reference-body let us imagine a spacious chest resembling a room with an observer inside who is equipped with apparatus. Gravitation naturally does not exist for this observer. He must fasten himself with strings to the floor, otherwise the slightest impact against the floor will cause him to rise slowly towards the ceiling of the room. (Theory of Relativity, Einstein)

Just as Einstein and others could predict the effects of near-zero gravity, we can predict that larger aquatic life lives outside earth.

That's right. It may seem like a bold statement, but it must be accurate. Why?

Let's analyze the first half of the statement. In 1916, Einstein could predict "weightlessness" in space. Briefly, he could predict that astronauts would float in the ship. Other scientists could do this as well based on classical mechanics. They didn't need to travel into space for physical confirmation.

Today, we can predict that larger aquatic life must live in places beyond the earth's atmosphere. We've investigated only eight planets. We've found at least one huge body of saltwater, Europa. Earth holds oceans. Let's ignore temporarily the other bodies that hold saltwater, Enceladus, Ganymede, and Callisto.

Opposite: Jupiter's moon Callisto. Image courtesy NASA.

The universe holds innumerable stars. If each star holds on average two large bodies of saltwater, the universe must hold countless bodies of saltwater. Let's take one trillion as a minimum estimate of the number of stars. Accordingly, I estimate the universe holds at least two trillion large bodies of saltwater. Of two trillion possibilities, larger aquatic life probably developed in one of these large bodies of saltwater.

Larger aquatic life must live in waters beyond earth.

I hope readers will forgive the redundancy, but this point is hardly accepted among all scientists. Einstein wrote the above passage in 1916. It's remarkable. Writing over four decades before humans traveled to outer space, he KNEW the effects of zero gravity. He wasn't the first. Many scientists understood.

That's why I recommend Einstein not only for the substance of his theories, but also his method. His meticulous method led him to profound insights. And if we adopt his careful method, we can at least gain a good understanding of physics.

Readers familiar with my other books OTHER LIFE EXISTS (2010) and KNOWN

(2011) know the truth. We KNOW that life lives in the waters of Jupiter's moon Europa. I know that. Thoughtful readers accept it. It is not necessary to sample the waters of Europa and confirm life under a microscope.

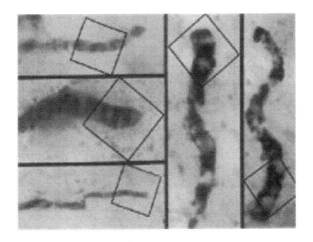

Microbial life. Image courtesy NASA.

Science gives us the ability to predict, and we know that life exists in Europa. We now know exactly where to find life. It's in Europa!

We have found life in every major body of saltwater on earth. And we know that geophysics applies to Europa. We've confirmed that Europa has an ocean beneath its icy surface. We used geophysics to confirm this fact. Electromagnetism, a field of geophysics, confirms Europa's ocean.

When the Voyager probe returned the first detailed images of Europa, scientists argued. What was beneath the surface ice? Did geophysics apply? What about Jupiter's immense radiation? How did that affect Europa?

The images showed ice that many believed resembled pack ice around earth's poles. Still, debate raged. The Galileo probe measured the electromagnetic fields around Europa. Everyone saw clearly! The field data confirmed that Europa must hold saltwater!

Of course, the scientists who simply relied on basic geophysics were not surprised.

Geophysics applies. So we know Europa resembles earth in some ways. We can equate the initial conditions for life between Europa's waters and earth's. Both are a mixture of sodium, hydrogen, and oxygen.

Europa holds a vast ocean. That ocean is over twice as big as all earth's oceans. It's incredulous to believe that Europa's waters are lifeless. We know of no examples of such a lifeless body of saltwater. Why presume one exists in Europa?

And the same logic that applies to Europa applies to Enceladus, the moon of Saturn. That moon harbors an ocean beneath its surface ice. The waters of Enceladus must hold life. And what will we find when we

explore the waters of Europa and
Enceladus?

God only knows.

The Galileo Probe explores Jupiter and its moons. Image courtesy of NASA.

Confirming Einstein

Einstein's general relativity has been confirmed countless times. Einstein elegantly discusses the development of science:

THE EXPERIMENTAL CONFIRMATION OF THE GENERAL THEORY OF RELATIVITY

From a systematic theoretical point of view, we may imagine the process of evolution of an empirical science to be a continuous process of induction. Theories are evolved and are expressed in short compass as statements of a large number of individual observations in the form of empirical laws, from which the general laws can be ascertained by comparison. Regarded in this way, the development of a science bears some resemblance to the compilation of a classified catalogue. It is, as it were, a purely empirical enterprise.

But this point of view by no means embraces the whole of the actual process ; for it slurs over the important part played by intuition and deductive thought in the development of an exact science. As soon as a science has emerged from its initial stages, theoretical advances are no longer achieved merely by a process of arrangement. Guided by empirical data, the investigator rather develops a system of thought which, in general, is built up logically from a small number of fundamental assumptions, the so-called axioms. We call such a system of thought a theory. The theory finds the justification for its existence in the fact that it correlates a large number of single observations, and it is just here that the "truth" of the theory lies.

Corresponding to the same complex of empirical data, there may be several theories, which differ from one another to a considerable extent. But as regards the deductions from the theories which are capable of being tested, the agreement between the theories may be so complete that it becomes difficult to find any deductions in which the two theories differ from each other. As an example, a case of general interest is available in the province of biology, in the Darwinian theory of the development of species by selection in the struggle for existence, and in the theory of development which is based on the hypothesis of the hereditary transmission of acquired characters.

We have another instance of far-reaching agreement between the deductions from two theories in Newtonian mechanics on the one hand, and the general theory of relativity on the other. This agreement goes so far, that up to the present we have been able to find only a few deductions from the general theory of relativity which are capable of investigation, and to which the physics of pre-relativity days does not also lead, and this despite the profound difference in the fundamental assumptions of the two theories. In what follows, we shall again consider these important deductions, and we shall also discuss the empirical evidence appertaining to them which has hitherto been obtained.
(Einstein, Theory of Relativity)

Dude. The universe holds larger aquatic life beyond earth. Definitely... Image courtesy of NASA, the NASA Earth Observatory, and Oceanic Institute.

Arthur Eddington first confirmed general relativity through astronomical observation. But in fact, Einstein confirmed himself even before Eddington's observations. Einstein knew from existing astronomical data that general relativity, not Newton's theory, best explained the orbit of Mercury.

Motion of the Perihelion of Mercury

According to Newtonian mechanics and
Newton's law of gravitation, a planet
which is revolving round the sun would
describe an ellipse round the latter, or,
more correctly, round the common centre
of gravity of the sun and the planet. In
such a system, the sun, or the common
centre of gravity, lies in one of the foci of
the orbital ellipse in such a manner that, in
the course of a planet-year, the distance
sun-planet grows from a minimum to a
maximum, and then decreases again to a
minimum. If instead of Newton's law we
insert a somewhat different law of
attraction into the calculation, we find
that, according to this new law, the motion
would still take place in such a manner
that the distance sun-planet exhibits
periodic variations; but in this case the
angle described by the line joining sun and
planet during such a period (from
perihelion--closest proximity to the sun--to
perihelion) would differ from 360 degrees.
The line of the orbit would not then be a
closed one but in the course of time it
would fill up an annular part of the orbital

plane, viz. between the circle of least and the circle of greatest distance of the planet from the sun.

According also to the general theory of relativity, which differs of course from the theory of Newton, a small variation from the Newton-Kepler motion of a planet in its orbit should take place, and in such away, that the angle described by the radius sun-planet between one perihelion and the next should exceed that corresponding to one complete revolution... (Theory of Relativity, Einstein)

To help general readers, I omit the math. "Perihelion of Mercury" means the closest point Mercury approaches the sun. Einstein's theory elegantly and gracefully explains Mercury's perihelion. He explains,

Our result may also be stated as follows : According to the general theory of relativity, the major axis of the ellipse rotates round the sun in the same sense as the orbital motion of the planet. Theory requires that this rotation should amount to 43 seconds of arc per century for the planet Mercury, but for the other Planets of our solar system its magnitude should be so small that it would necessarily escape detection.

In point of fact, astronomers have found that the theory of Newton does not suffice to calculate the observed motion of Mercury with an exactness corresponding to that of the delicacy of observation attainable at the present time. (Theory of Relativity, Einstein)

Eddington and his team confirmed Einstein during a 1919 solar eclipse. Einstein continues:

(b) Deflection of Light by a Gravitational Field

In Section 22 it has been already

mentioned that according to the general theory of relativity, a ray of light will experience a curvature of its path when passing through a gravitational field, this curvature being similar to that experienced by the path of a body which is projected through a gravitational field. As a result of this theory, we should expect that a ray of light which is passing close to a heavenly body would be deviated towards the latter...

This result admits of an experimental test by means of the photographic registration of stars during a total eclipse of the sun. The only reason why we must wait for a total eclipse is because at every other time the atmosphere is so strongly illuminated by the light from the sun that the stars situated near the sun's disc are invisible...

In practice, the question is tested in the following way. The stars in the

Enceladus with tiger stripes. Image courtesy
of NASA and Cassini Team.

neighbourhood of the sun are
photographed during a solar eclipse. In
addition, a second photograph of the same
stars is taken when the sun is situated at
another position in the sky, i.e. a few
months earlier or later. As compared with
the standard photograph, the positions of
the stars on the eclipse-photograph ought
to appear displaced radially outwards
(away from the centre of the sun) by an

amount corresponding to the angle a.

We are indebted to the [British] Royal Society and to the Royal Astronomical Society for the investigation of this important deduction. Undaunted by the [first world] war and by difficulties of both a material and a psychological nature aroused by the war, these societies equipped two expeditions--to Sobral (Brazil), and to the island of Principe (West Africa)--and sent several of Britain's most celebrated astronomers (Eddington, Cottingham, Crommelin, Davidson), in order to obtain photographs of the solar eclipse of 29th May, 1919. (Einstein, Theory of Relativity)

So Einstein predicted that light from a distant star would deflect around the sun during a solar eclipse. Eddington's team proved him right.

The gravitational field of the sun actually "bends" the path of the light from a distant star.

Astronomers call this effect "gravitational lensing." A distant star emits light. On the way to earth, the light passes a second star. The second star's gravity pulls the light and deflects its path.

Author's Note

I wrote this while battling electronic
hacking. Some call it electronic warfare.
Whatever the name, it will be the end of
the Internet.

It was bad enough during the days of the
stand-alone microcomputer. A bug
intentionally placed on a disk might move
from computer to computer. Writers were
at wit's end trying to stop misprints. I think
the problem is even worse on the web.

I try. I hope readers will excuse an
occasional misprint caused by hacking.

INDEX

Wade Hobbs graduated from George Washington University, where he studied with scientists from the National Aeronautics and Space Administration. He worked five years in the patent field. USA Today, National Public Radio, and the American Institute of Physics have carried his work. He spent a year in the U.S. Coast Guard Auxiliary.

He is in the vanguard of researchers proving that life exists outside Earth.

Made in the USA
Charleston, SC
20 August 2014